Walk Around

By William Scarborough
Color by Don Greer
Illustrated by Joe Sewell

PBY Catalina

Walk Around Number 5

squadron/signal publications

Introduction

The intent of the WALK AROUND series is to provide maximum information in a limited number of pages. This led us to the decision to restrict this book to the PBY-5/5A/6A and their variants, USAF OA-1OA and Catalinas/Cansos of the Allied Air Forces.

Considered obsolete by many as the Second World War II began, the PBY proved a tremendous asset to all the services that flew her, performing in roles never envisioned by her designers. Maritime patrol, convoy escort and ASW were her primary duties, but in all combat areas, especially the South Pacific, the PBY proved to be a capable bomber and outstanding Search and Rescue Dumbo, saving hundreds of survivors. Bombing and strafing "Black Cats", overall Black PBYs flying at night, were the scourge of enemy efforts to supply bypassed forward area bases. Basically unchanged from first production in 1936 to the last PBY, delivered in 1945, the Catalina's operational capability had increased enormously with improved armament, power-boosted ammunition supply, armor, fuel dump valves, self-sealing fuel tanks, thermal deicing, radar and communication gear.

PBYs continued to serve many armed services after the war. The last Navy Catalina, a PBY-6A, flew until 1957 in the Naval Reserve. Foreign air forces operated PBYs into the 1970s, principally for search and rescue and logistic support of outlying installations.

Catalinas have served in many civilian roles since the Second World War, providing passenger and freight service to remote areas and for many more exotic pursuits. Some serve as water bombers, fighting forest fires all over the World with water scooped into hull tanks through retractable probes as the PBY skimmed over a lake or river surface. Several have been configured as air yachts by private owners. For improved performance, many have been re-engined with 1,700 hp Wright R-2600 engines and nacelles from B-25 bombers. A revised vertical tail improves stability and control and has led to a new name - "Super Cat".

Acknowledgements

Photos and documentation were provided by the individuals listed below and by U.S. Navy (USN), National Archives (NA), U.S. Naval Institute (USNI), and Consolidated Aircraft Corp. (CAC):

Hal Andrews	Bob Lawson	Mark Aldrich
Ross Creed	Dive Lucabaugh	Richard Dann
Dave Menard	Fred Dickey	Bill Mewhorter
John Elliott	Jim Mooney	Jeff Ethell
Brown Ryle	Bill Hardaker	Frank Strnad
Bill Larkins	Mike Bobe	Tom Tullis

Members of the Flying Boat Amateur Radio Society
And apologies to those I have missed!

Personnel of the Naval History Center and the Smithsonian NASM Aeronautics Department provided material and access to files which was invaluable, with special thanks to Hal Andrews and Bill Hardaker at NASM. Major sources of color photos were the restored PBY-5 at the National Museum of Naval Aviation at Pensacola (NMNA), the PBY-5A at the San Diego Aerospace Museum (SDAM), the OA-10A at Dayton's USAF Museum (USAFM) and the PBY-6A of the Confederate Air Force, Southern Minnesota Wing.

COPYRIGHT 1995 SQUADRON/SIGNAL PUBLICATIONS, INC.
1115 CROWLEY DRIVE CARROLLTON, TEXAS 75011-5010
All rights reserved. No part of this publication may be reproduced, stored in a retrieval system or transmitted in any form by means electrical, mechanical or otherwise, without written permission of the publisher.

ISBN 0-89747-357-4

If you have any photographs of aircraft, armor, soldiers or ships of any nation, particularly wartime snapshots, why not share them with us and help make Squadron/Signal's books all the more interesting and complete in the future. Any photograph sent to us will be copied and the original returned. The donor will be fully credited for any photos used. Please send them to:

Squadron/Signal Publications, Inc.
1115 Crowley Drive
Carrollton, TX 75011-5010

Если у вас есть фотографии самолётов, вооружения, солдат или кораблей любой страны, особенно, снимки времён войны, поделитесь с нами и помогите сделать новые книги издательства Эскадрон/Сигнал еще интереснее. Мы переснимем ваши фотографии и вернём оригиналы. Имена приславших снимки будут сопровождать все опубликованные фотографии. Пожалуйста, присылайте фотографии по адресу:

Squadron/Signal Publications, Inc.
1115 Crowley Drive
Carrollton, TX 75011-5010

軍用機、装甲車両、兵士、軍艦などの写真を所持しておられる方はいらっしゃいませんか？どの国のものでも結構です。作戦中に撮影されたものが特に良いのです。Squadron/Signal社の出版する刊行物において、このような写真は内容を一層充実し、興味深くすることができます。当方にお送り頂いた写真は、複写の後お返しいたします。出版物中に写真を使用した場合には、必ず提供者のお名前を明記させて頂きます。お写真は下記にご送付ください。

Squadron/Signal Publications, Inc.
1115 Crowley Drive
Carrollton, TX 75011-5010

Overleaf: A PBY-5A on the dirt ramp of a forward base in the Aleutian Islands during December of 1943. (Via Jeff Ethell)

PBY-5A **PBY-5**

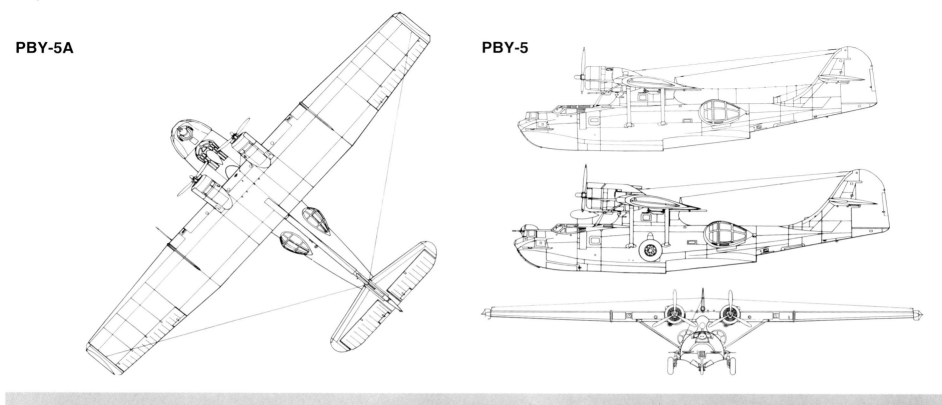

A trio of PBY-5s of Patrol Wing 4 (PatWing 4) over the Aleutian Islands during early 1943. The lead aircraft was 43-P-8, while the second Cat was 42-P-2.

The restored PBY-5A Catalina at the San Diego Aerospace Museum is mounted on a pylon in the museum center court. From above, there are a number of details that are rarely seen including the wing tip antenna posts, wing tip lights, and tops of the engine nacelles. This PBY-5A carries a teardrop Automatic Direction Finder (ADF) antenna mounted on the upper wing between the engines. (Mark Aldrich)

The side windows in the cockpit canopy opened and slid backward on tracks in the fuselage side for opening. The chine rail, with its mooring cleats, was used by the crew to stand on while anchoring or mooring the PBY to a buoy. The two upper canopy windows also opened, sliding to the rear. The small U shaped panel in the upper right side of the insignia bar is a pull-out step, used by the crew for climbing up on the fuselage from the chine rail. (Mark Aldrich)

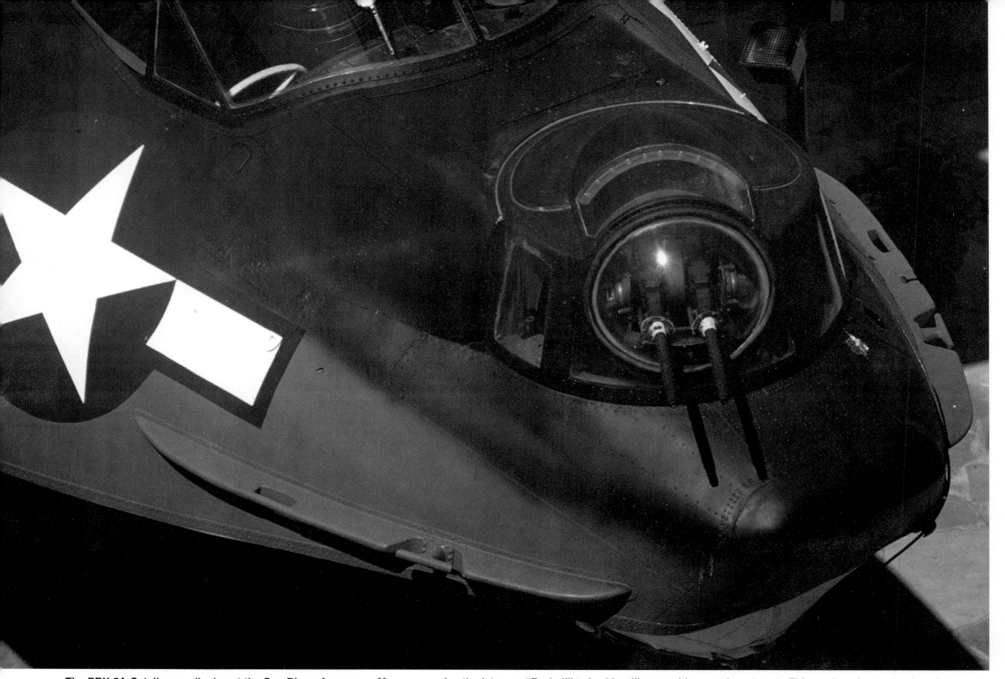

The PBY-5A Catalina on display at the San Diego Aerospace Museum carries the late war "Eyeball" twin .30 caliber machine gun bow turret. This enclosed turret replaced the open single gun position used on earlier PBY variants. Other features visible include the chine rail with its mooring cleats. (Mark Aldrich)

The upper canopy of the PBY-5A featured rear sliding windows on both sides that doubled as upper fuselage escape hatches. The PBY used a wheel mounted on a control yoke for aileron and elevator control There were windshield wipers installed on both sides of the windscreen to assist in clearing away spray during water takeoffs and landings as well as in rain and during deicing. The side windows on both sides of the cockpit also opened, sliding to the rear. (Mark Aldrich)

The PBY-5A at the San Diego Aerospace Museum is mounted on a pylon in the center Court. Visible are the cockpit, wing support pylon which housed a mechanic's station, wing center section and engine nacelles. An ADF teardrop antenna housing and antenna lead-in mast for the flat top antenna was mounted above the wing center section. (Richard Dann)

The starboard bow area of the SDAM BY-5A. From left: fuselage ice shield, which was a metal doubler to prevent damage from ice thrown from the propellers during icing conditions. The small vent at the bottom corner of the windshield can be cranked open from inside the cockpit. A pull out step is located at the upper right corner of the insignia. (Richard Dann)

The port bow area of the SDAM PBY-5A reveals the windshield and wipers, bow gun turret, chine rail (used by the crew when anchoring and mooring), anchor stowage compartment door, nose wheel compartment doors and, to the rear of the door, a vent for water drainage after a water takeoff. (Richard Dann

The overhead hatches on the PBY-5A opened to the rear. This aircraft was a civil owned PBY-5A with extra electronics and antennas, (Ross Creed)

The windshield of the PBY was curved and equipped with wind shield wipers to help remove rain, ice and the sea spray during water operations, (T. Tullis)

The port side window just behind the cockpit was located in the radio operator/navigator station.. (Milke Aldrich)

The side windows in the cockpit slid open to the rear. The instruments visible behind the pilots seat are the electrical control panel. (B. Ryle)

The knob visible on the side window is the handle for moving the sliding window. There are windshield wipers on both sides of the windscreen. (Mike Aldrich)

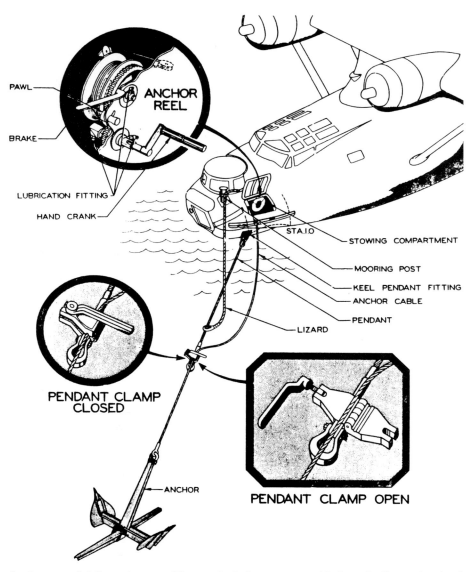

Anchor gear details and usage. The pendant clamp was used to transfer the anchor load from the anchor cable to the bow pendant, which was permanently attached to the keel. The Lizard line was used to retrieve the pendant when reeling in the anchor cable. The mooring post was removable and was not always used. The detachable hand crank was used for the anchor, for manually cranking engines and for lowering or raising the wing tip floats when electric power not available. (USN Manual)

The anchor was stowed in a compartment located on the port side of the bow near the bow turret of all PBY variants. The PBY-5 and later models carried a smaller anchor of the same type. Operating instructions were carried on the inside of the compartment door. (CAC)

The bow gun turret of a early production PBY-5 during 1940/41. The turret was armed with a single .30 caliber air cooled machine gun. Belted ammunition for the gun was in 100 round magazines (ammo cans) in the rack visible under the gun. The socket below the gun turret was used for mounting the removable snubbing post. (CAC)

This bow turret was installed on a USAAF OA-10 during 1942/43. The turret was still armed with a single Browning .30 caliber machine gun, but the mounting was modified to provide some armor protection for the gunner. This same turret was also used on the PBY-5 and PBY-5A. (CAC via Lucabaugh)

(Right) This is the bow turret compartment of a PBY-5A Catalina during 1943, with the machine gun and other gear stowed for a ferry flight. The perforated cans on the side of the gun were for spent cartridges and links, the square one for links and the other for cartridges. The object below the gun is a bow hook with a rope used to help secure the Catalina when mooring to a buoy. The crewman stood on chine rail to catch a ring on top of buoy with the hook as the aircraft taxied up to it, paying-out line, around the bow snubbing post to stop the aircraft gradually after catching the buoy. Bow equipment (from lower left) includes intervalometer for releasing bombs in train, bomb rack circuit breakers, bomb rack and fuze selector switches (top left), manual arming and release handles on left and right side of the bombing window. Panel (left center top) had bombing circuits. (USN)

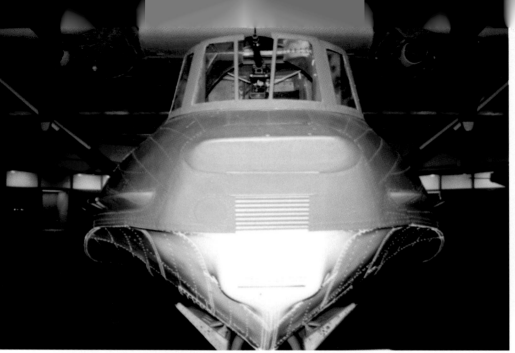

The nose of an OA-10A (PBY-5A). The nose wheel doors are visible at the bottom. The bombing window in the center is covered over by a retractable shutter for protection. On either side of the hull are the chine rails, which are used as a work platform by crewmen when mooring the Catalina. This aircraft has an early style bow turret with a single .30 caliber machine gun. (B. Ryle)

This is the starboard side of the bow turret of the USAF Museum's OA-10A. The gun mount used on this restored aircraft was not the same as that used in wartime PBYs, although the gun itself is the correct type used. (B. Ryle)

The port side of the bow turret of an OA-10A. The gun is held in place with a bungee cord for display purposes. The Browning .30 caliber air cooled machine gun was the standard armament used in the bow turret of all early PBYs. (B. Ryle)

While the Browning machine gun and bow turret are correct, the gun mount used in the restored OA-10A at the USAF Museum is not of the type actually used in the PBY. The bow turret had fixed windows plus a removable center section and top hatch. (B. Ryle)

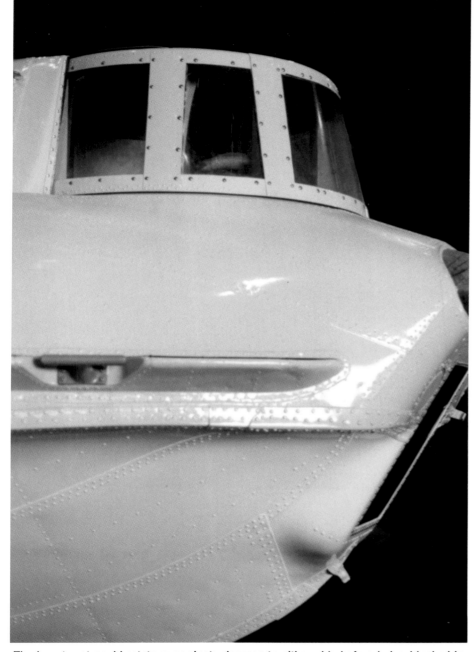

The bow turret could rotate over ninety degrees to either side before being blocked by the fuselage. The front center panel of the turret, through which the gun barrel normally extended for firing, could be covered with a removable panel whenever the gun was in the stowed position. (Mark Aldrich)

Later PBY-5As had a twin .30 caliber machine gun turret which became known as the "eyeball" turret. This turret greatly improved the gunner's ability to train and aim the twin guns when firing. (Richard Dann)

The "eyeball" bow turret was a big improvment over the earlier open turret. In addition to improving accuracy the new turret kept the gunner out of the elements and slipstream. The vertical post just below the turret is the snubbing post, which was removable and usually stowed when not in use. (Mike Bobe)

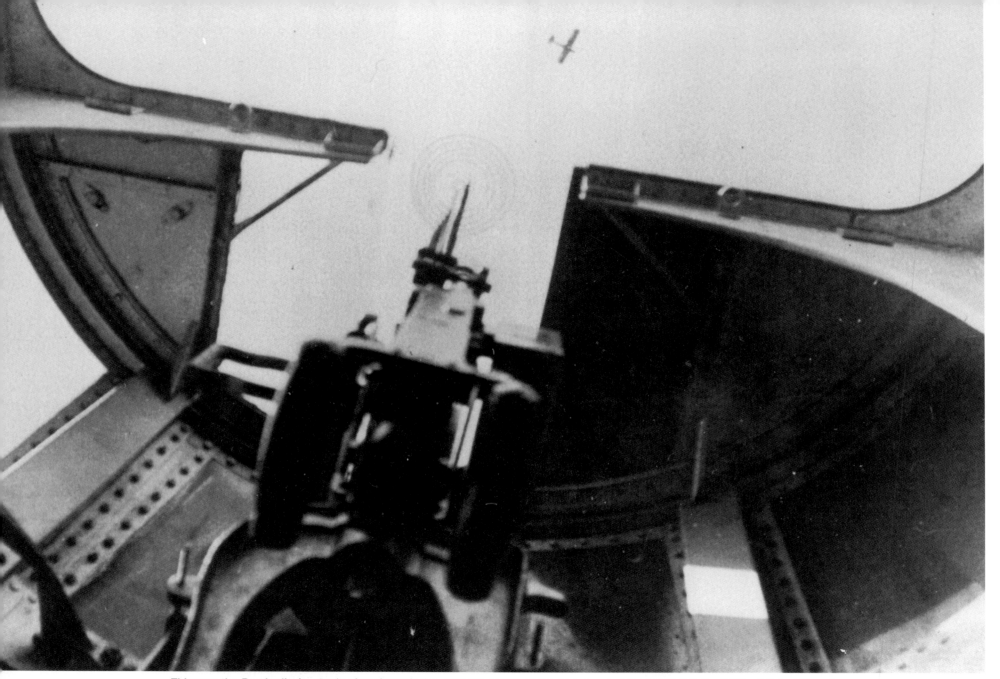
This was the Bombadier/gunner's view through the open top hatch of the bow gun turret of PBY-5 BuNo 2350, of Patrol Squadron Seventy-two (VP-72) during September of 1941. (WES)

A PBY-5 of an Operational Training Squadron based at Naval Air Station Jacksonville, Florida, during March of 1943. This aircraft is equipped with the early style bow turret. (USN via Bob Lawson)

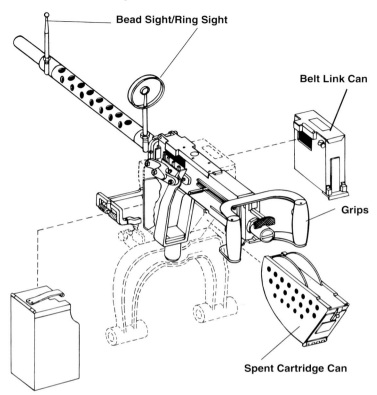

Early Bow Turret Gun Installation

- Bead Sight/Ring Sight
- Belt Link Can
- Grips
- Spent Cartridge Can
- Ammunition Can

This PBY-5 (BuNo 2292) of VP-52 was transferred from VP-14 during January of 1941. The rudder was repainted in White, VP-52's assigned color. VP-52 was stationed at NAS Norfolk, Virginia, during March of 1941. (WES)

This "eyeball" .30 caliber machine gun turret was removed from a PBY and placed on display The "eyeball" turret had two additional Plexiglas sections, the gun mount sphere and the clear upper dome cover. This mount was a modification of the earlier bow turret and retained the lower turret portion with its multiple windows. This turret was carried by all late production PBY variants and was retrofitted to many earlier aircraft as they went through rework. (Mark Aldrich)

The Plexiglas sphere holding the twin .30 caliber machine gun mount swiveled in elevation with the guns as a single unit, like an "eyeball" within a socket. The gunner stood within the bow compartment with his head inside the Plexiglas dome behind the guns. There were two large ammunition cans mounted just below the gun, bolted to the turret ring which rotated with the turret. (Mike Bobe)

This PBY-5A is in long term open storage at NAS Pensacola, Florida, for future restoration and display. This Catalina has had the bow turret modified at some point in its career, with the gunner's dome being replaced by a sheet metal fairing. The round object at lower right is the top of the removable mooring post. (Mike Bobe)

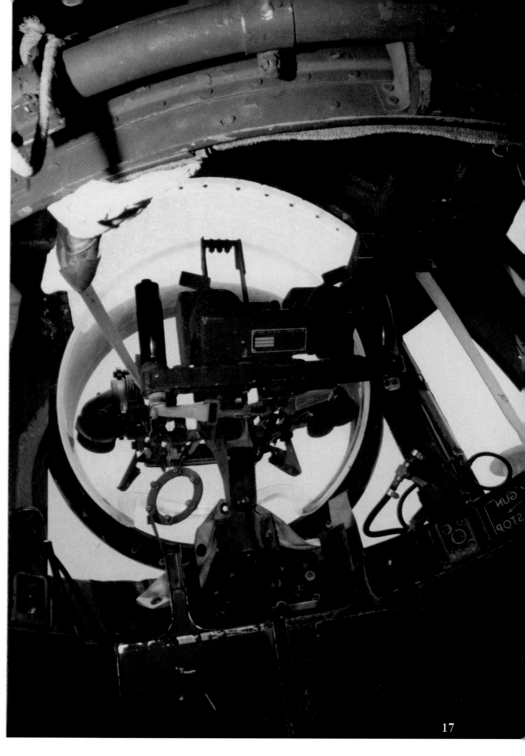

The interior of the bow turret on the stored PBY still retained the twin .30 caliber machine gun mount, twin ammunition cans (2,100 rounds total) mounts, Mk 9 reflector gun sight mount and twin gun trigger handles. The interior of the bow compartment was Chromate Green with the gun equipment in Black. (Mike Bobe)

Defensive Armament and Armor Installations, PBY-5/5A

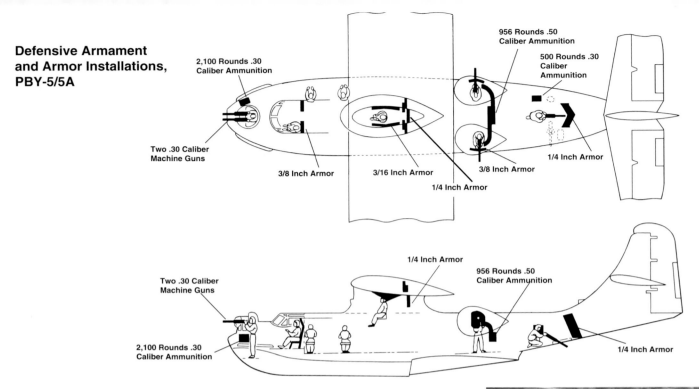

Twin .30 Caliber Machine Gun "Eyeball" Bow Turret

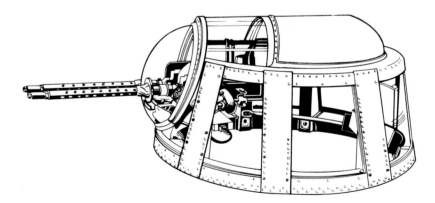

(Right) The "eyeball" turret modification was highly successful and well liked by PBY crews. It improved the bow guns handling, doubled the firepower of the position, and kept the gunner out of the weather. (Mike Bobe)

This PBY-5A, stationed on Guam during August of 1945, was equipped with the tunnel gun blister mount, which is visible just to the rear of the waist positions and forward of the tail. (via Mark Aldrich)

Tunnel Gun Blister Mount.

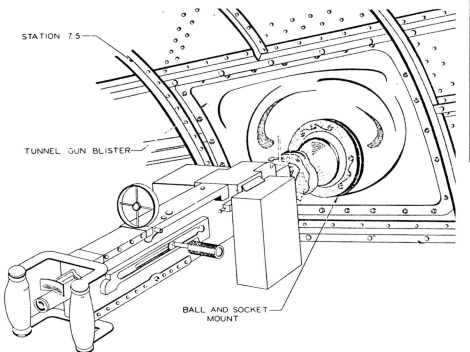

Tunnel gun position with the hatch closed and the gun in the firing position. The ammunition can is on the left side, belt link container on the right and spent cartridge container below gun. An armor shield is to the rear of the hatch. The White bag is ballast carried while engaged in a ferry flight with less than a full crew. (USN)

The two mounting sockets for tail beaching gear struts were located on the bottom of the rear hull. (WES)

Removing and reattaching beaching gear posed many of the least desirable aspects of seaplane operations. For launching and recovering the P-boats, the U.S. Navy developed over the years many operating techniques based primarily on the use of beaching crews. For PBYs, crews generally had 6 or 7 men, two for each side mount, one at the tail to steer with the tail wheel tiller while it was moving on beaching gear, a safety watch ahead and a crew chief in charge. For launching, the crew followed the plane down the ramp into the water controlling its speed by the tractor tow line and by the side mount men with the brakes. Once afloat the Beachmaster signalled the men on the side mounts to pull the retaining pins. On release, the gear floated up and the men walked it back up the ramp. The man on the tail gear unpinned it and walked it up the ramp, When all were clear, the Beachmaster released the tow line and signalled the aircraft's crew to taxi clear of the ramp. Beaching a PBY was a far more demanding operation, especially in high wind and sea states. Many years of experience in handling seaplanes and flying boats produced techniques and equipment which made safe beaching practical but, unfortunately, space here is too limited to discuss them. Basically, all were developed to position the aircraft on the ramp and hold it there until the beaching gear was attached. (WES)

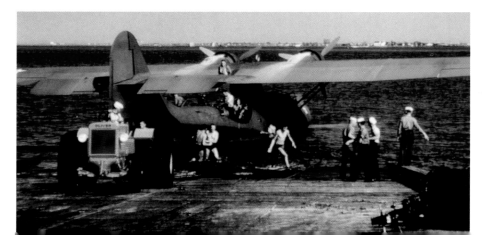

The tail beaching gear mounted on the hull of the PBY-5 at the Naval Aviation Musuem at Pensacola, Flordia, was up on tempoary blocks. (WES)

The beaching crew moves a SPBY-5 (BuNo 08318) of the Training Command up the seaplane ramp at Bronson Field, Pensacola, Florida, during 1946. The prefix "S" in the designator on the tail has not been identified. The beach crewman at the aircraft's tail steered the PBY, using the beaching gear tiller as it was towed up the ramp. (P. Fackler via R.

The survivors of PatWing Ten (VP-10) from the Phillipines reorganized and re-equipped with new PBYs in Australia. These VP-101 PBY-5s are parked on their beaching gear at an advance base at Pelican Point, Perth, Australia, during August of 1942. (USN/NA via Dave Lucabaugh)

For ferry flights between bases, the main beaching gear and tail wheel were disassembled and stowed in the bunk compartment. The gear was plainly marked with right hand and left hand markings and tied down securely before takeoff. (USN)

A PBY-5 of VP-101 at Pelican Point, Perth, Australia, prepares to launch from an improvised ramp. Once in the water the beach crew will remove the beaching gear and stow it on the beach, like the set in the foreground. (USN/NA via Dave Lucabaugh)

The starboard engine nacelle of a PBY-5A with the cowl flaps closed. The intake on the lower nacelle is the oil cooler housing with coolant vents on each side. The logo on the propeller blades is the company logo for the Hamilton Standard company. (Richard Dann)

The port engine nacelle of the San Diego Aerospace Museum's PBY-5A. The tube like object above the nacelle is one of the two exhaust stacks. There is a single auxiliary air intake on this side of the nacelle just behind the cowl flaps. (Richard Dann)

An engine and propeller of the USAF Museum's OA-10A. The Catalina used Pratt & Whitney R-1830-92 radial engine and a three blade Hamilton Standard Hudromatic full feathering propeller. (B.Ryle)

The lower rear starboard engine nacelle of a USAF OA-10A. The oil cooler honeycomb is visible and there are several vents long either side for additional cooling. The cowl flaps are partially open. The small vent in the center is a oil drain valve. (B. Ryle)

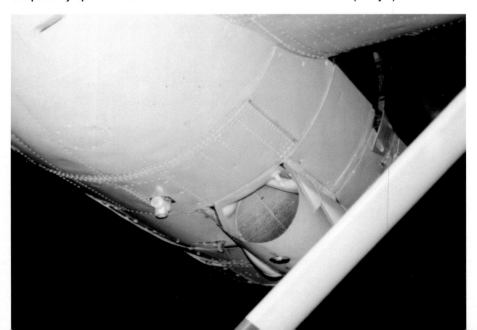

This PBY-6A Catalina is owned by the Confederate Air Force and is powered by Pratt & Whitney R-1830-92 air-cooled radial engines driving twelve foot, three blade Hamilton-Standard constant-speed propellers. The work stands attached into sockets above the wing and in the forward portion of the cowling, allowing maintenance personnel access to the engine. The work stands were normally carried aboard service PBYs stowed in the bunk compartment. (Mark Aldrich)

(Above) Aviation Machinist Mate 1c E. F. Dyer used the portable work stands that were supplied with each aircraft to work on the engine. When working at a buoy, a tarpaulin was spread under the nacelle between the work stands to catch any dropped parts or tools. (WES)

(Left) E.F. (Pappy) Dyer, AMM1, USN, was the Plane Captain (Crew Chief) of 52-P-9, a PBY-5 (BuNo 2293) assigned to Patrol Squadron Fifty-Two (VP-52). He uses a socket set and portable work stand to change the engine's spark plugs at Naval Air Station Norfolk, Virginia, during February or March of 1941. The author was a NAP, (enlisted pilot) assigned to Dyer's crew. The wire behind Dyer was a support wire that ran from the top to the bottom of each work stand. (WES)

PBY engine work stands in place on each side of the engine nacelle. The V strut above the propeller was a portable propeller hoist used to change propellers in the field. These, like the work stands, were the same for all PBY variants. (USN)

(Right) The propeller hoist was a piece of squadron maintenance equipment and was not usually carried in the aircraft. Two crew men from the "Prop Shop" maintenance crew guide the propeller as it was lowered. The propeller would then be put on a dolly and moved to a shop for maintenance work. (USN)

The wire support line was bolted to the upper and lower work stand vertical posts. This PBY has oil stains on the cowl flaps and under the engine nacelle, not an unusual sight on all PBYs. (Mark Aldrich)

Maintenance personnel use portable work stands to check the engine of a PBY-5 equipped with a thermal wing deicing system shroud on the port engine exhaust. The cowl flaps are in the full open position. (Mark Aldrich)

A PBY-5, 11-P-7 of VP-11 is hoisted aboard the seaplane tender USS TANGIER riding at anchor in the bay at Noumea, New Caledonia, on 8 March 1942. In addition to VP-11, VP-71 and VP-72 deployed detachments to Noumea during the build-up to the battle of the Coral Sea. The object resting on the hull to the rear of the waist blisters was a sea anchor drying out after use. These were used to control taxi speed and direction on the water. (National Archives via D. Lucabaugh)

A crewman stands by with a fire extinguisher as the pilot prepares to start engines of this PBY-5, No. 72, in July of 1943, in the Aleutian Islands. Once started, the aircraft will be taxied to its assigned buoy in the seaplane anchorage. The crewman in the bow turret is standing by to release the mooring line from the snubbing post when signaled by the pilot. Lines hanging from the port blister are attached to a sea anchor used to control speed during approach to the ship. This Catalina is equipped with flame damper exhaust stacks and a thermal wing deicing system. (USN/National Archives via Dave Lucabaugh)

Crewmen wash down a PBY-5 of VP-52 (52-P-1) with fresh water after a mission. Removal of salt water helped cut down on corrosion and prolonged the life of the airframe. This aircraft carries the early style U.S. national markings in use prior to 1942 with the Red center dot. The wing support pylon on the fuselage center also housed the flight engineers station and there was a window on each side of the pylon. (National Archives via Mark Aldrich)

The wing center support pylon was the main support for the massive 104 foot wing. The pylon housed the flight engineer's station with his control panels and equipment. There was a window on each side of the pylon. The ledge half way up the front of the pylon was the pylon step used by personnel to gain access to the top of the wing from the fuselage. This PBY has two non-standard antenna masts on the upper wing along with a pair of non-standard hand holds, probably added by the CAF. (Mark Aldrich)

The wing center support pylon of a PBY-5A. The small window allowed the flight engineer some view outside the aircraft. to visually check each engine There was an identical window on the other side of the pylon. There is a small non-standard hand hold below the window. (Mark Aldrich).

The two wing support struts were secured to the wing with a single bolt that ran through an eye that was attached to a main wing spar. (Mark Aldrich)

Aircraft in service had the strut attachment bolt points covered by a fairing to cut down on drag. (Mark Aldrich)

The fuselage strut attachment points were on either side of the main landing gear wheel wells on the PBY-5A and later variants. These struts were each attached to a main fuselage structural frame on all PBY variants. (Mark Aldrich)

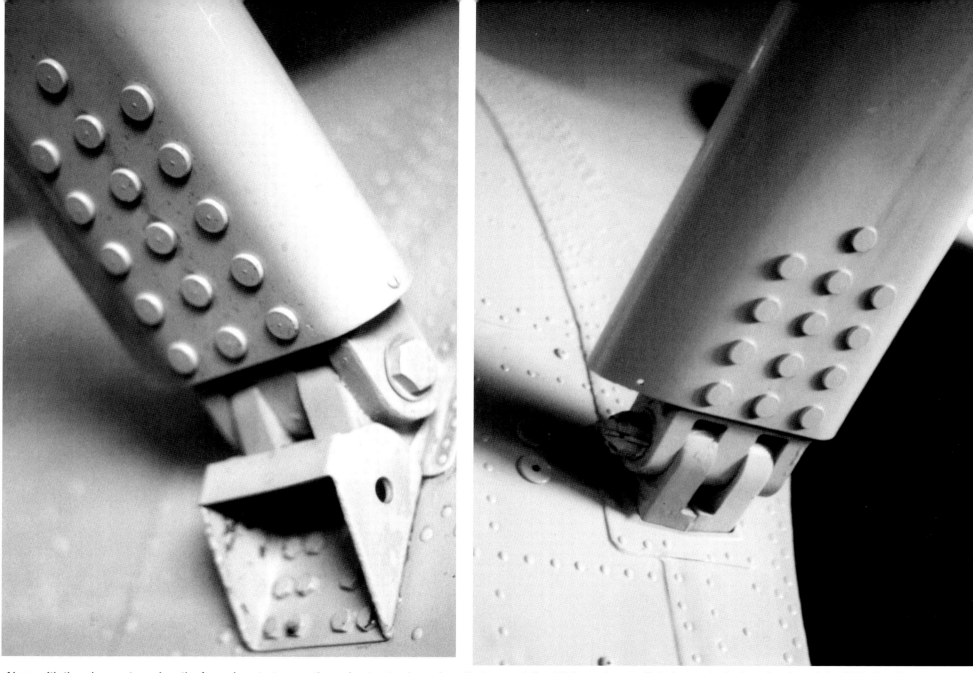

Along with the wing center pylon, the four wing struts were the main structural members that support the 104 foot wing. In flight in rough air, the wing tips of the PBY often flexed upward and downward. The attachment points were mounted on two of the main fuselage frames located on either side of the main landing gear wheel well of the PBY-5A. The forward frame formed the aft bulkhead for the radio/navigation compartment and the rear frame formed the forward bulkhead of the bunk compartment. (Mark Aldrich)

The retractable wing tip floats on the PBY-5A at the San Diego Aerospace Museum are in the fully retracted position. The rings on the front and rear of the float was tie down rings. The fairing at the wing leading edge just inboard of the float is the starboard position light, and the post near the wing trailing edge was the wing tip antenna mast. An antenna wire ran from this mast back to the forward leading edge of the vertical fin, while another wire ran from the fin to the opposite wing tip. (Mark Aldrich)

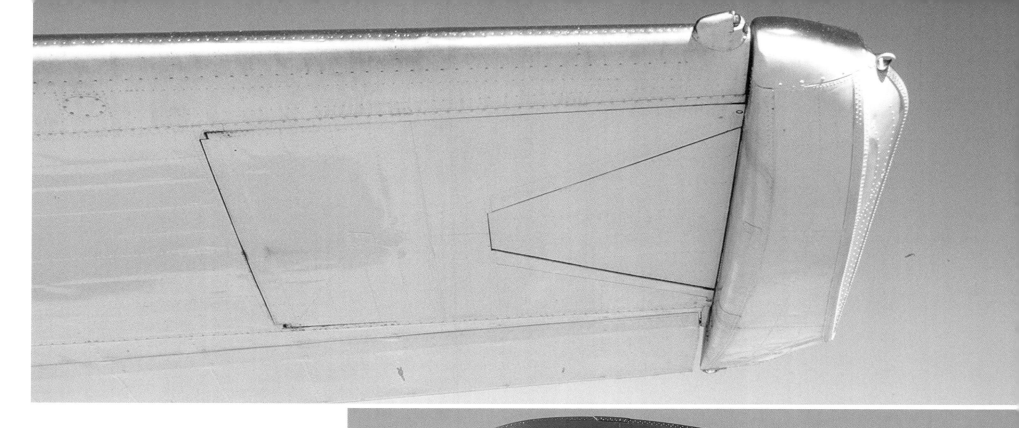

When retracted, the wing tip floats of the PBY fit flush with the tip of the wing while the doors fit flush with the underside of the wing. The V shaped section is a structural part of the lower wing surface and not a part of the float. The wing tip navigation light shines forward and outward and is shielded from the cockpit by the light fairing. The ring on the float front is a tie-down ring used to help secure the aircraft. When retracted, the float is held in position by the fully retracted actuating strut. The color demarcation line runs down the center of the float, with areas view from above painted in the upper surface camouflage color and areas viewed from below painted in the lower surface color. The aileron trim tab is also visible along with its actuating push rod. In rough air the floats could be observed flexing up and down. (Mark Aldrich)

This PBY-5, beached on the Naval Air Station Pensacola, Florida, shows one way to change crews between training flights. The wing tip float is in the full down position and is resting on the sand. The X shaped struts are the float actuating and support struts.

Normally, when operating from land bases, the wing tip floats were retracted for takeoff and flight, but were lowered to serve as post flight tie-downs. The strut coming back to the wing from the X float support struts was the float retraction linkage strut. When retracting, the float supports folded inward at the cross of the "X". This is a Canadian-built Vickers Canso A.

This PBY-5B, is backed up to the beach at NAS Pensacola, Florida, between flights. The two notches in the wing tip were the wells for the float support struts. (Fred C. Dickey)

During the take-off run, the wing tip floats did not touch the water once the aircraft was on the step. In fact, if the aircraft was balanced, they rarely touched the water, except in heavy seas or when turning during water taxi. (CAC)

This restored PBY-5A is on display at Naval Air Station Jacksonville, Florida. The wing tip floats are in the full down position. The strut coming back off the strut support struts is the retraction linkage which retracts into the well within the float well. The float support struts flod inward lying in the notches in the wing tip. The Jacksonville PBY was modified with the bow turret being removed and faired over. The aircraft was painted with the wrong colors and markings. In addition, the tail codes are presented wrong and the fuselage codes are presented wrong for this type of paint scheme. (Richard Dann)

An alternative to beaching the aircraft on a ramp was operating from a suitable beach, especially for short duration training flights. NAS Pensacola routinely operated both primary and advance patrol plane training from beaches from before the First World War until after the Second World War. Crewmen in "boots" (anti-exposure suits) are attaching handling lines to the wing tip float tie-downs to be used to reposition the aircraft for engine start and launch for the next flight. (W. Derby via J. Ethell)

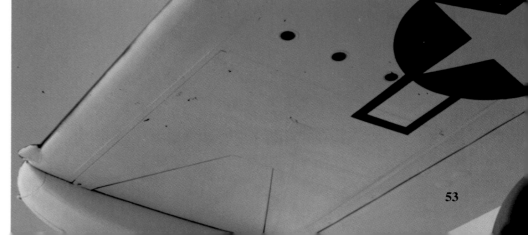

The starboard wing of a PBY-5A showing the three underwing recognition lights that were mounted just inboard of the retractable wing tip float. The lights were(from back to front) Red, Green and Blue. These lights were used for identification, displaying one of more lights in color patterns ordered by the area operational commander for specific time periods both day and night. (Richard Dann)

Side by side it is easy to see the differences in the fin and rudder area of the PBY-5A (left) and PBY-6A (Right). The PBY-6A had a three and a half foot taller fin and rudder. This tail modification was derived from engineering and wind tunnel studies which led to superior control responses on the PBN-1, these modifications were then incorporated into the PBY-6. The rudder balance forward of the hinge line on the PBY-5 was a post war civil modification intended to improve handling and reduce control forces. . (Mark Aldrich)

This PBY-5A at NAS Patuxent River, Maryland was equipped with the following radio gear in the radio/nav compartment. From left at the forward end of the compartment: GO-9 Transmitter with ATB Command Transmitter mounted above it. Antenna reel is to right. On the shelf above the radioman's desk are RV-19 Receiver, LM-10 Frequency Meter and ARB Receiver. Above them are Antenna Transfer and Clock panel and ICS Control Box. At upper right, mounted on aircraft structure, are two round bright end objects which are the Main Electric Power Distribution Panel. Below it are four coils which allow transmitter frequency changes. Continuing forward: radioman's desk with a storage locker at aft end with life jacket stowage pocket on its door and radioman's chair. On desk, under shelf, is the transmitter key, ZA Impedance Adapter, ARB Control Box and RU-19 Junction Box (bright connector on bottom. Below desk are ISC Dynamotor and Demolition switch. See pages 64/65 for additional details of radio gear. (USN)

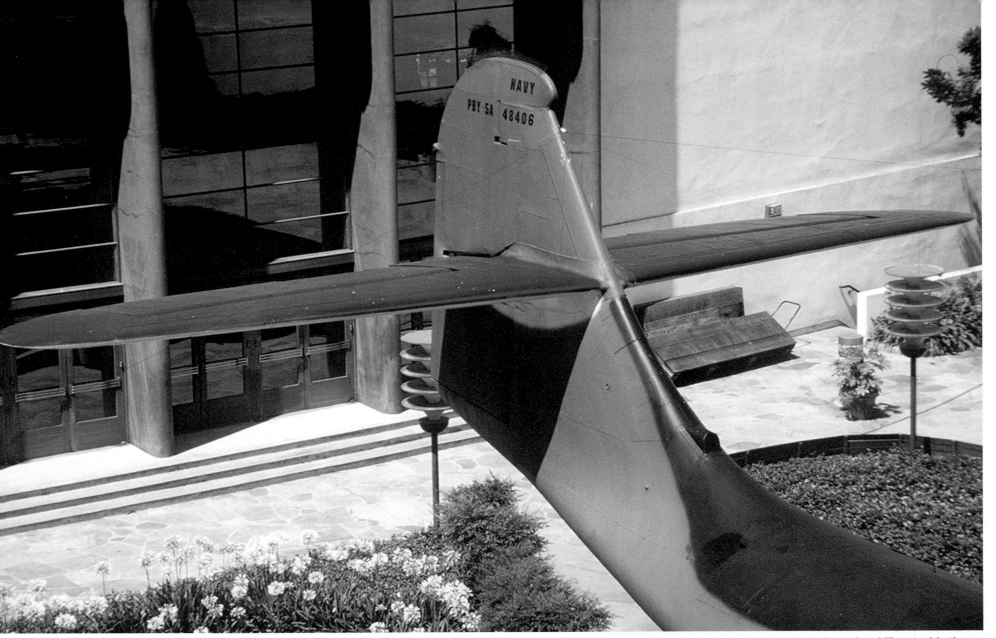

The vertical and horizontal stabilizers and control surfaces of the PBY-5A at the San Diego Aerospace Museum. The elevators are inset in the horizontal stabilizer and both are equipped with trim tabs. The round fairing just above the rudder trim tab is the tail navigation light. The small white object just below the tip of the fin on the forward edge is an antenna wire insulator. At least three antenna wires are attached to this insulator. The protrusion at the base of the fin is the air intake for the thremal deicing system. The rudder balance forward of the rudder hinge line reduced control forces and was not found on service PBY-5/5As. It was probably a modification installed by the previous civil owner. (Mark Aldrich)

The fin and rudder of the PBY-5 on display at the National Museum of Naval Aviation. The trim tab actuator was only on this side. Both elevators were also equipped with trim tabs. (WES)

The vertical and horizontal tail, control surfaces and trim tabs of the San Diego Aerospace Museum's PBY-5A. The rudder control cable and control horn are visible just below the horizontal tail. (Richard Dann)

The opposite side of Navigator/Radio compartment. From left, forward: box for Navigation books with mount for Pelorus on side, Navigator's chair, Navigation table with drawer for chronometers (two were visible through the round windows in table), stowage for Octant and Binoculars. Compass was located on table at aft end. The meter at top and additional instruments above table are test equipment, standard Navigation instruments are above test equipment and include Temperature Gage (top) Airspeed, and Altimeter. The large drawer is for chart stowage. The semi-circular tray above Octant Stowage is a drip pan under hydraulic actuators for the auto pilot. The cone at top is the Navigator's Flood Light. (USN)

The navigation/radio compartment crew at work aboard "Wandering Wench" a Royal Canadian Air Force Canso, on 4 September 1943. The radio operator was at the left, the Navigator at right. The other crewman is adjusting the main electric power distribution panel. (Public Archives of Canada)

The vertical fin, rudder and port horizontal stabilizer of the restored PBY-6A that is operated by the Southern Minnesota Wing of the Confederate Air Force. The PBY-6A had a taller fin and rudder than the earlier PBY-5A, and this particular aircraft has been further modified with modern electronics antennas on the leading edge of the fin. The wicks hanging down from the elevator are static dischargers, which bleed off static electricity from the aircraft. (Mark Aldrich)

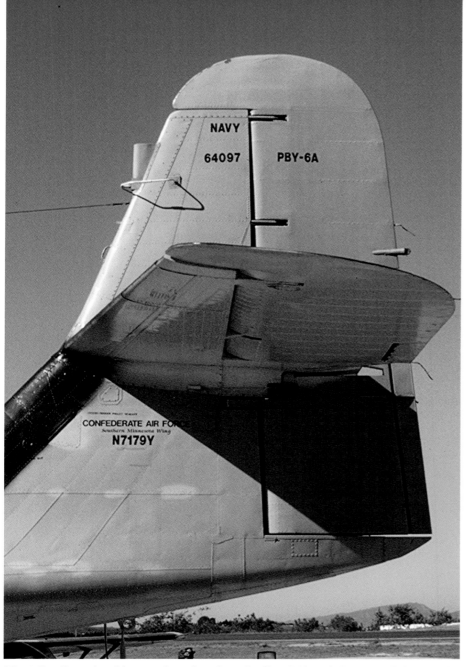

The taller fin and rudder was introduced on the Naval Aircraft Factory PBN-1 and the Navy specificd it for production PBY-6As. Other than the modiried tail, the aircraft was basically a PBY-5A. (Mark Aldrich)

A tie-down rope is attached to the rear fuselage tie-down eye under the fin and rudder of this PBY-6A. Tie-down eyes were located under the nose, tail, leading edge of the outer wing panels and on each of the wing tip floats. (Mark Aldrich)

The Mechanic's compartment looking forward and up. At the lower left was the APU fuel tank. Control cables to the tailare on each side of the Mechanic's seat. The cushion was for the protection of crew members walking under the seat. (USN

Left forward corner of the Mechanic's compartment: At left was the multi-purpose crank used for manual operation (if power failed) of the engine starter, anchor reel, and wing tip floats. The fuel tank for the auxiliary power unit (APU) was mounted above the APU. At the top is the bottom of Mechanic's seat, which was ;located within the wing support pylon and, at right,were the steps to the seat. (USN)

The mechanic's compartment was aft of the Nav/Rdo Compartment. The three steps led to the Mechanic's seat located in the wing support pylon. Aft of the steps was an electric stove with two water breakers (tanks) above it. The housing at the right was for the landing gear strut. (USN)

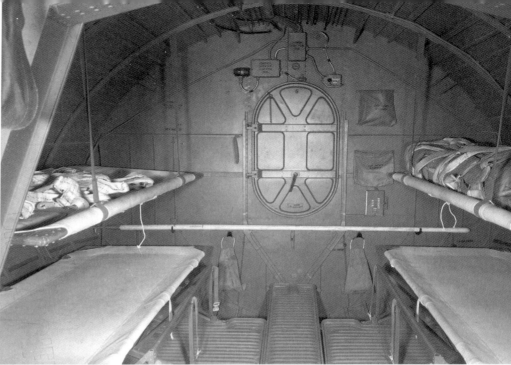

This was the bunk compartmen of a PBY-5 (BuNo 48262) at NAS Patuxent River, Maryland during 1943. The bulkhead at the rear of the compartment is Bulkhead 6. Bunks are located on both sides of the compartment. The frames and web straps in front of the bunks are for stowing Portable Work Stands. A parachute harness is on the left top bunk. A boat hook was stowed in hooks on the bulkhead. The slings located at the lower bulkhead are stowage for two parachutes. At the right of the bulkhead door were stowage for two life jackets with a First Aid Kit being stored below them. Parachutes are visible on the right hand bunk. (USN)

(Left) Looking forward from the Navigation/Radio compartment through the door from the Mechanic's Compartment. The hump in deck provides clearance for the nose wheel when retracted. Inspection window was for checking the condition of the nose wheel. The backs of the pilots seat are visible, to the left is the Plane Commander (pilot), while the right seat was for the copilot. The two "T" handles located on the back of the seats were for adjusting the seat position. (USN)

The reel with the handle is the trailing antenna. On the shelf above the desk is a RU-19 radio receiver, LM-10 frequency meter and Radio Operator's ARB unit. Above these are an antenna transfer and clock panel and ICS control. The radio operators sending key is on the desk. (Jim Mooney)

The starboard side of the Navigator/Radio compartment of the PBY-5 at the National Museum of Naval Aviation at Pensacola. From left: door to cockpit, the Black box next to the door is a long range GO-9 radio transmitter. The two boxes above the GO-9 are ARB (left).and ATB (right) Command Receiver & Transmitter. (Jim Mooney)

(Right) Located at the forward end of the compartment was the base of the long range GO-9 radio transmitter and its mounting posts. (Jim Mooney)

The right rear of the Navigation/Radio compartment of the PBY-5 at the National Museum of Naval Aviation at Pensacola. From forward end, on shelf above radio operator's desk: LM-10 frequency meter and ARB receiver. On a shelf above the LM-10 and ARB is the ISC control. On bulkhead, main power distribution panel. At top two engine generator voltage regulators. (Jim Mooney)

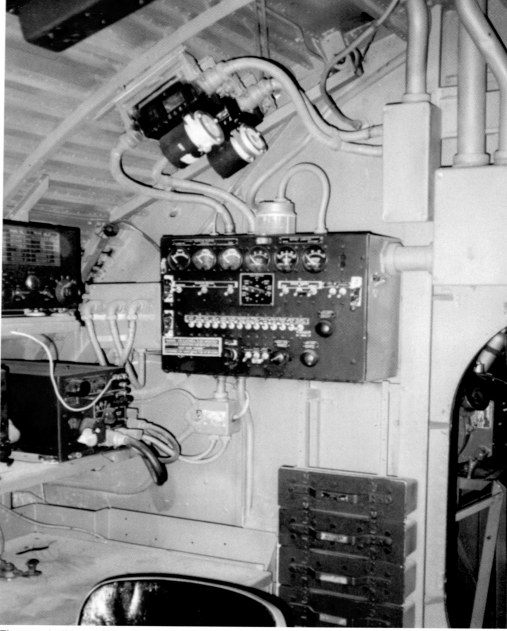

The rear bulkhead of the Navigator/Radio compartment. From lower left: Radioman's chair, Radio Key on desk, ARB Receiver, Antenna Transfer and Clock Panel, Engine Generator Voltage Regulators. On the aft bulkhead, Main Power Distribution Panel, below it are two coils for changing frequencies. The door is to the flight engineer's compartment. (Jim Mooney)

These are the pilot's rudder pedals on a PBY-5 flying boat. The round-top lever on the left side of each rudder pedal is the pedal lock, used to adjust the pedals for different pilots. (Author)

This was the cockpit of a PBY-5 flying boat. On the left side of the instrument panel was the Pilot Director Instrument (PDI), for bombing runs under the direction of the Bombardier. The controls, yoke and manual bomb releases were the same as for the PBY-5A. There was no brake pedal mounted with the rudder pedals. Located in the bow compartment was the Norden Mk 15 bombsight which was mounted in the center above the leather knee pad. The fabric curtain under the instrument panel is a draft curtain which could be closed and zipped shut to stop drafts from the bow compartment entering the cockpit. The track on top of glare shield is the mount for Torpedo Director, used when armed with torpedos. (USN)

(Above) Looking forward into the cockpit of a PBY-5A. The pilot's rudder and brake pedals are at the bottom left, the pilot's control wheel is above, mounted on the cross cockpit yoke. On top of the yoke, was the signal system between the pilot's and mechanic. On the right end of the signal box are the controls for the deicing boots on the wing and tail. Under the yoke is the ignition switch for the engines and controls for wing-mounted recognition lights. Below the yoke and instrument panel are two "T" handles for manual salvo or emergency release of the wing rack bombs. Identifiable instruments visible on the left panel (from left, top row) Instrument Landing System, Gyro Compass, Clock. Middle row: Air Speed, Turn and Bank, Rate of Climb. Bottom row, Altimeter, Directional Gyro, Gyro Horizon. On center panel from left, top row, Tachometers, Manifold Pressure and Synchroscope indicator and on right, the Floats up warning Light. The Sperry Auto Pilot was behind the Signal Lights. Right panel top row: Radio Altimeter, Gyro Compass. Middle row: Blank, Turn and Bank, Rate of Climb. Bottom row: Altimeter and Air Speed indicators. Below the panel are the Copilot's Rudder and Brake Pedals. The cockpit was painted in Zinc Chromate Green with Black instrument panels, yoke and control wheels. (USN)

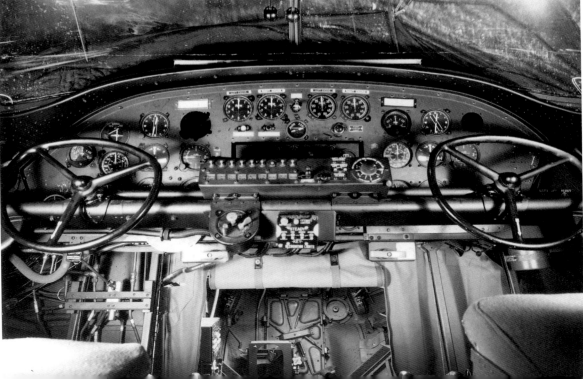

The panel installed at the top of the bulkhead behind the pilot's seat contained the controls for Intercom and Radio, on the top panel and the lighting controls (interior and exterior) on the lower panel. The receptacle for the remote "Pickle Switch" for bomb and torpedo release was located to the left of the lighting panel. (Author)

This is the bulkhead behind the pilot seats. The Black panel on the bulkhead has, from top down, two Black "T" handles which are pulled to release parachute flares from flare tubes in the tunnel compartment. Tuning control on the right side of the panel is for voice and radio range receiver, used for navigation and instrument flight. Toggles and knobs on the lower panel are radio controls. Placards in center of the panel identify combinations of radio receivers and intercom for each pilot. (Mooney)

The throttle and propeller controls were located overhead between the pilots. (Mike Bobe)

This is the engine control panel in the cockpit of the Confederate Air Force PBY-6A. The switches at the left are for engine start and include the booster pump, primer and starter switches for each engine. The float retraction switch is the Red switch in the center of the panel, next to it are the cowl flap switches and carburetor heat switches. The round switch below the panel is the main engine ignition controls. These modifications were required by the deletion of the engineer's station in the plyon. (Mark Aldrich)

A PBY-5A on the ramp at Randolph Field, TX during October of 1942. The aircraft was Blue Gray over Light Gray with Black numbers and codes. (via J. Ethell)

This PBY-5A has been modified for the transport role with the bow turret and waist positions deleted and faired over. The civil owned Catalina was on the ramp at Kirkland Air Force Base during 1986. (via Dave Menard)

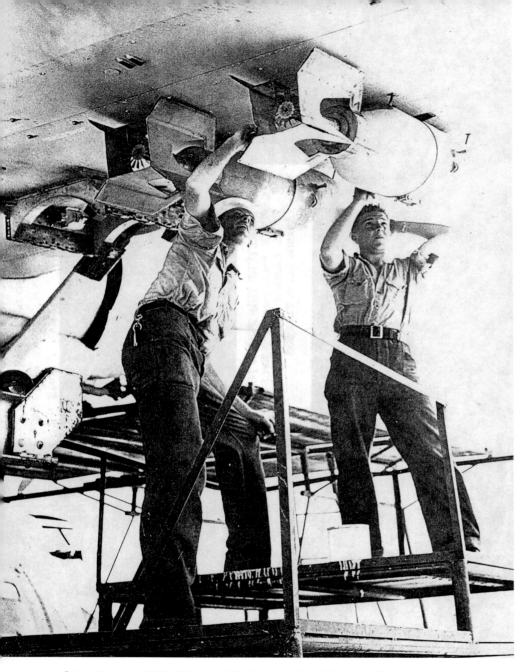

Ground crews of this PBY-5A of VP-91 are loading 500 pound General Purpose bombs on 5 June of 1942, just prior to the Midway battle. (National Archives)

These crewmen are loading 500 pound General Purpose bombs on the underwing racks of a PBY-5A under harsh conditions at Cold Bay, Alaska on 23 November 1942. The antenna above the crewman at the right is a Yagi radar antenna. (George E. Earle)

Crews transfer 500 pound General Purpose bombs from a weapons carrier to the underwing racks of a PBY-5A of VP-91 on Midway Island on 5 June 1942. A torpedo rack is installed between the bomb rack and wing strut. (National Archives)

A bomb cart load of Mk 17 depth bombs is moved into position for loading on a PBY-5 of VP-72 at NAS Kaneohe, Hawaii during the Spring of 1942. (W. J. Henning, PH2 USN)

Another underwing load that could be carried by PBYs was an AR-8 dropable lifeboat. By 1950, the USCG had adopted new colors and markings. A USCG PBY-5A with White AR-8 dropable lifeboat. The aircraft is overall Aluminum with USCG emblem on side of hull below the cockpit on both sides. There is a Orange Yellow band with Black piping around the rear hull, The Catalina had Orange Yellow wing tips (top and bottom) and post war style national insignia. (USCG photo via Bob Lawson)

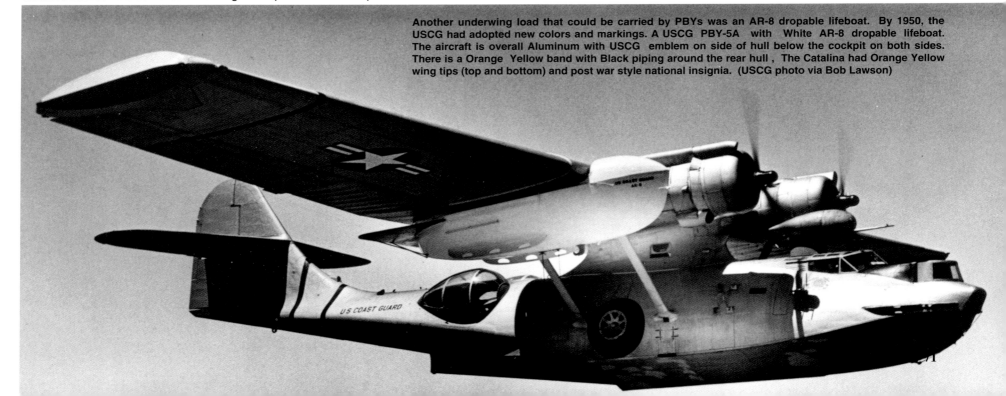

An Air Force OA-10A (43-3999) parked in the snow at Goose Bay, Labrador. The aircraft carries Yellow markings on the fuselage, wing tips, cowl tops and a large aircraft number in Yellow on the fuselage underside. (E. Van Houten via D. Menard)

This OA-10 (44-33939) carries both underwing radar antennas and a radar housing over the cockpit. In addition to its high visibility Yellow marking, this Air Force search and rescue aircraft also has a Red tail, required for Arctic operations. (W. T. Larkins)

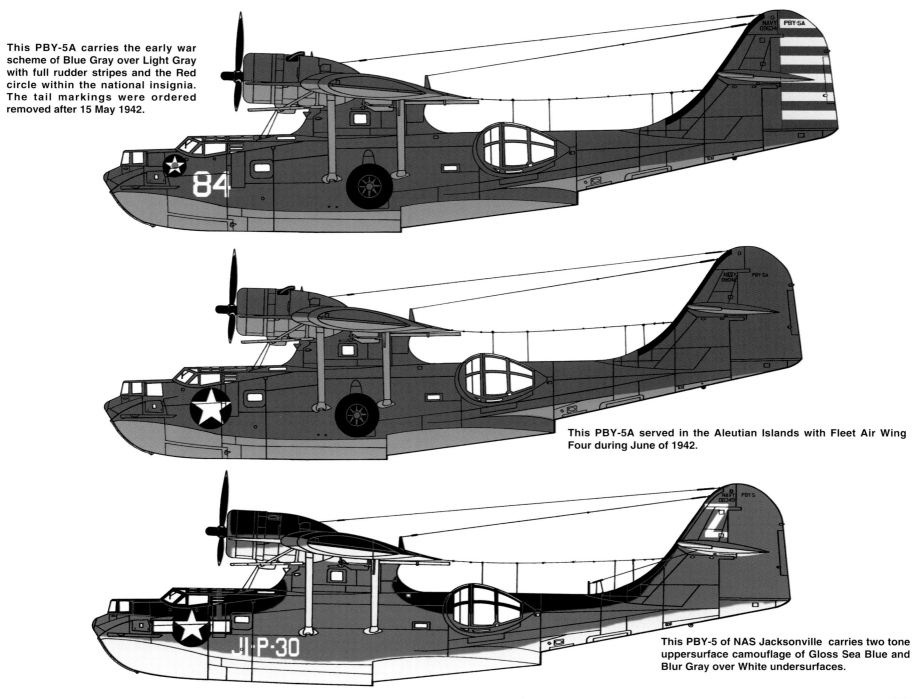

This PBY-5A carries the early war scheme of Blue Gray over Light Gray with full rudder stripes and the Red circle within the national insignia. The tail markings were ordered removed after 15 May 1942.

This PBY-5A served in the Aleutian Islands with Fleet Air Wing Four during June of 1942.

This PBY-5 of NAS Jacksonville carries two tone uppersurface camouflage of Gloss Sea Blue and Blur Gray over White undersurfaces.

PBY-5A, side number 18 of VP-63 during 1944, is fitted with tail mounted MAD gear. The aircraft is armed with anti-submarine retro rockets under the wings and has exhaust shrouds fitted for thermal wing deicing. The aircraft was painted in the anti-submarine scheme of Sea Gray upper surfaces over White. (USN

In December of 1943, VP-63 moved to Port Lyautey in French Morocco to fly a barrier patrol across the Straits of Gibraltar, utilizing MAD gear installed in the tail of their PBYs. The barrier proved effective and VP-63 was credited with sinking U-761 on 2/24/44, U-392 on 3/16/44, and U-731 on 5/15/44. In January of 1945 a detachment returned to Pembroke Dock, England and commenced patrols in the English Channel. On 30 April, the squadron was credited with a fourth U-boat, sinking U-1055. During May and June all aircraft were flown back to the States to prepare for decommissioning (USN)

Anti-submarine retro rockets under the port wing of a PBY. Retro rockets were developed by the California Institute of Technology (CIT) during the Second World War. The 65 pound 2.25 inch rockets carried a contact-fused explosive charge which avoided alerting a submarine to the presence of the aircraft if the rocket missed. A direct hit could produce significant damage. (USN)

This PBY-5 of VP-63, side code "O", was flown by Crew 15, from Pembroke Dock, Wales in September of 1943, on anti-submarine patrols armed with MAD gear in the tail, radar antennas on the wings and retro rockets underwing. (H. Lee)

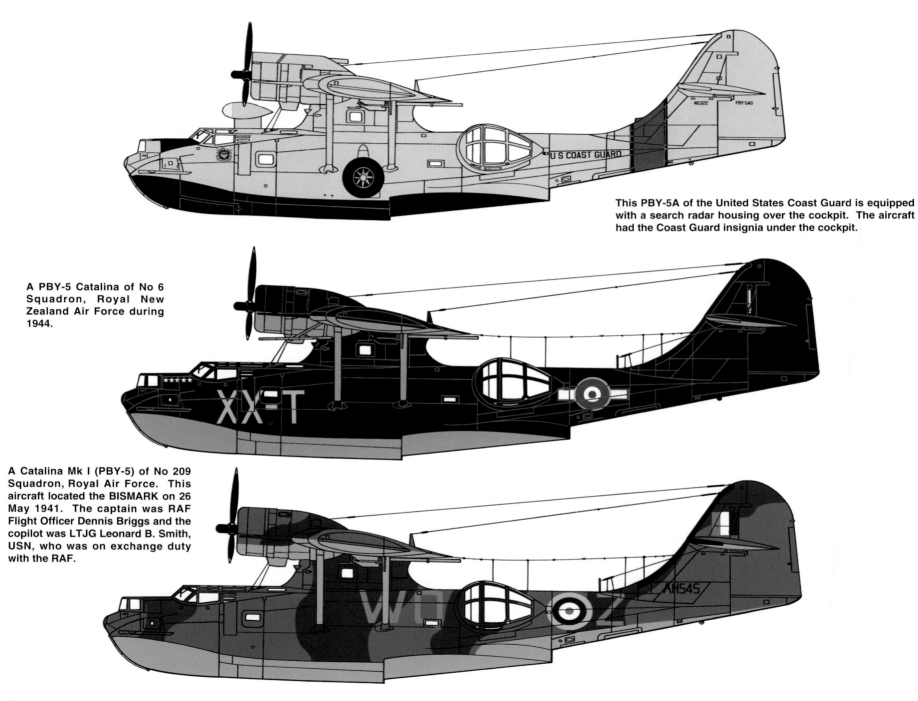

This PBY-5A of the United States Coast Guard is equipped with a search radar housing over the cockpit. The aircraft had the Coast Guard insignia under the cockpit.

A PBY-5 Catalina of No 6 Squadron, Royal New Zealand Air Force during 1944.

A Catalina Mk I (PBY-5) of No 209 Squadron, Royal Air Force. This aircraft located the BISMARK on 26 May 1941. The captain was RAF Flight Officer Dennis Briggs and the copilot was LTJG Leonard B. Smith, USN, who was on exchange duty with the RAF.

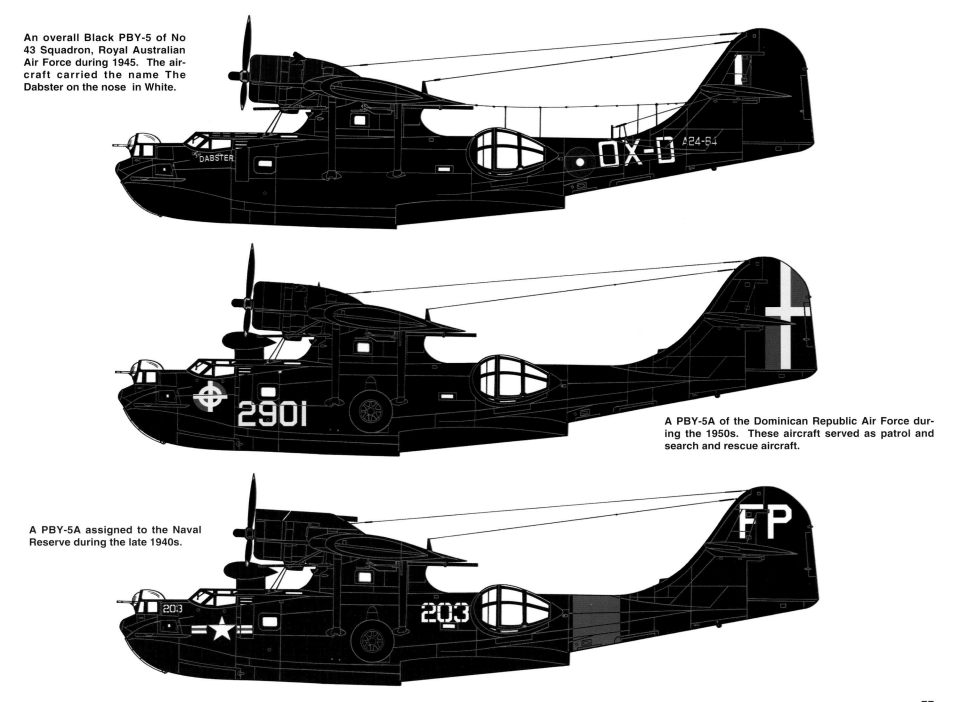

An overall Black PBY-5 of No 43 Squadron, Royal Australian Air Force during 1945. The aircraft carried the name The Dabster on the nose in White.

A PBY-5A of the Dominican Republic Air Force during the 1950s. These aircraft served as patrol and search and rescue aircraft.

A PBY-5A assigned to the Naval Reserve during the late 1940s.

The starboard Jet Assisted Take Off (JATO) bottle installation on a PBY-5A (BuNo 04406) on 2 June 1945. The aircraft could carry eight JATO bottles, four on each side and these could be fired either all at once, or in pairs (forward and aft). (USN)

A USCG PBY-5A demonstrates water takeoff with JATO assist at USCGAS Miami, Florida, during September of 1949. The Coast Guard operated PBYs from 1941 to 1945, maintaining an average 114 aircraft on board until 1945. Operating levels were gradually reduced after 1945 and, by 1954, all PBYs had been replaced by more modern aircraft. (USCG)

A PBY-5 Black Cat of VP-34 was shore-based at Samarai, on 3 March 1944. It was outfitted with an experimental fixed nose gun battery of four .50 caliber guns. Its typical bomb load was made up of 500 pound bombs hung outboard and 1,000 pound bombs hung inboard. (National Archives via Dave Lucabaugh)

BuNo 1245, was the last PBY-4, modified by Consolidated and redesignated XPBY-5A when delivered in December of 1939. Records show it at Anacostia in early 1940, then at Norfolk as a Staff aircraft for PatWingsLant until transfer to Patwing 3 at Coco Solo in late 1942. It was transferred to the Assembly & Repair Shops in March 1943 and was modified at that time to the transport configuration. When or where the landing gear was removed and the hull rebuilt is unknown. It was transferred to Pensacola, apparently to the training command and finally stricken in 1946.

This Black Cat PBY-5A in the South West Pacific on 7 April 1944, carried a field installation of two fixed forward firing 20mm cannons mounted in the bow compartment. (Dave Lucabaugh)

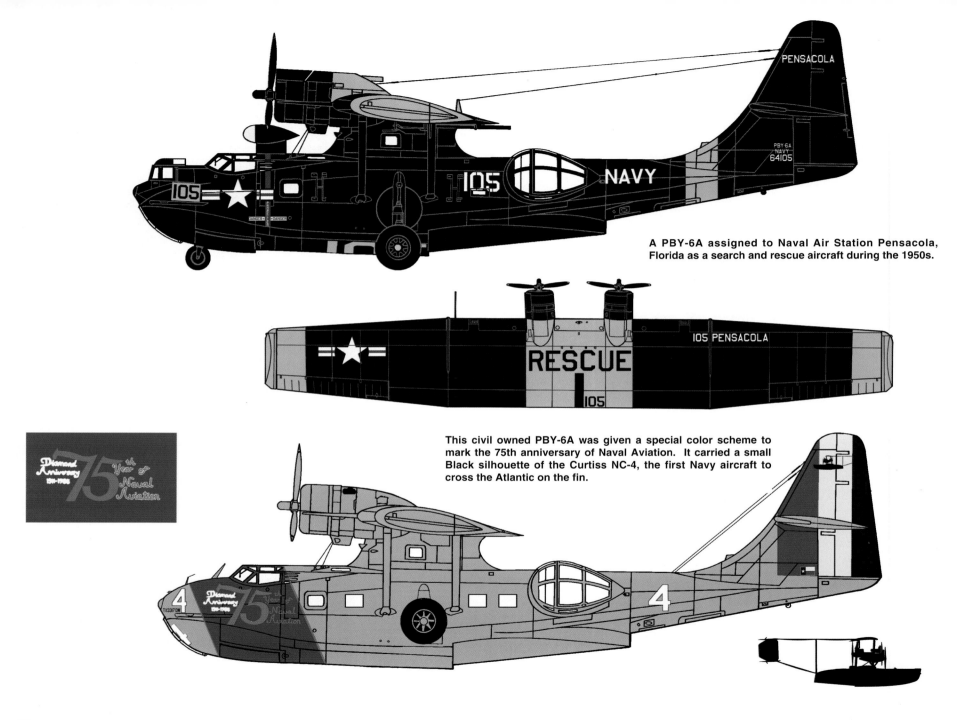

A PBY-6A assigned to Naval Air Station Pensacola, Florida as a search and rescue aircraft during the 1950s.

This civil owned PBY-6A was given a special color scheme to mark the 75th anniversary of Naval Aviation. It carried a small Black silhouette of the Curtiss NC-4, the first Navy aircraft to cross the Atlantic on the fin.